AF333174

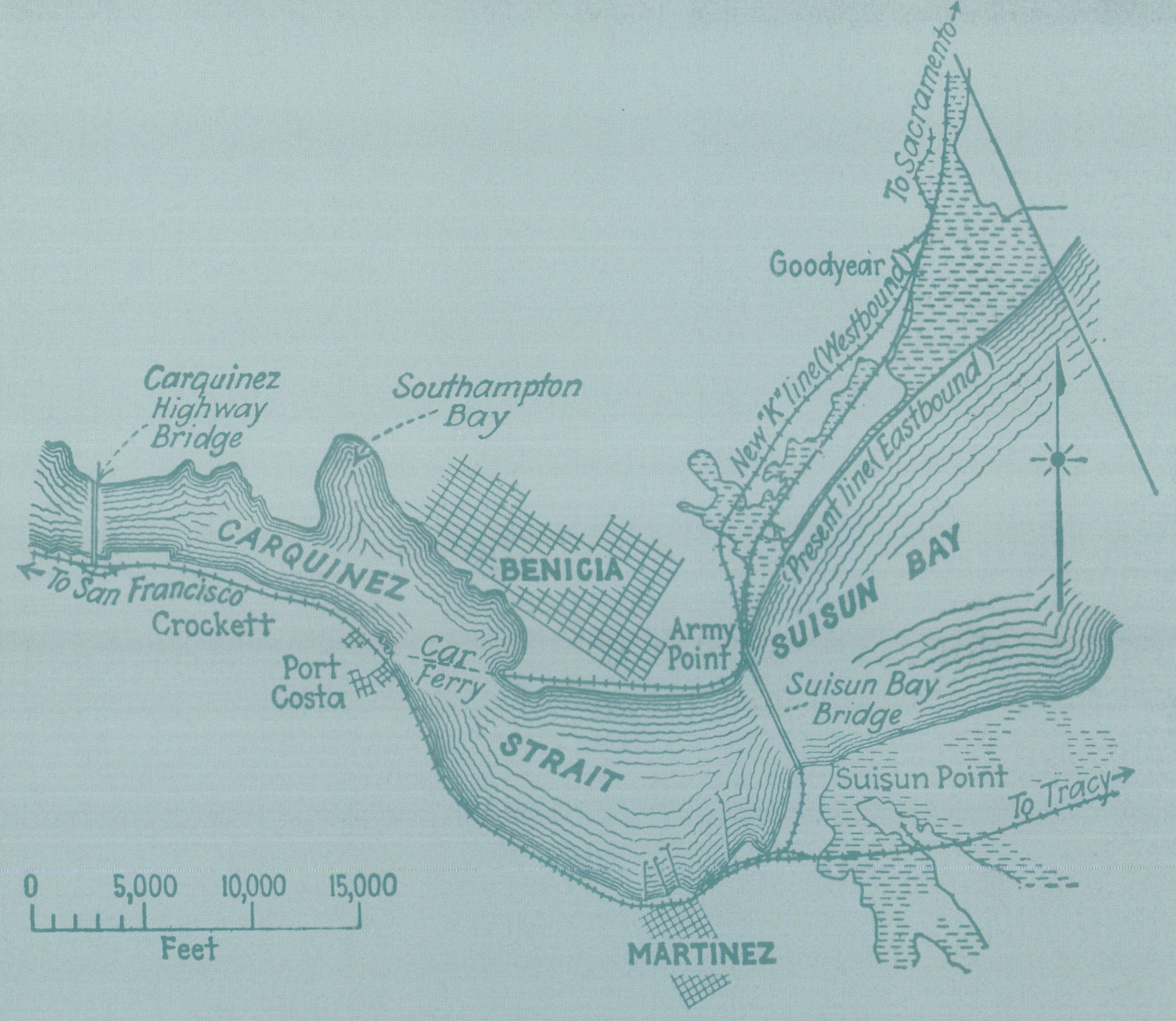

LOCATION MAP. NOTE CAR FERRY IN CARQUINEZ STRAIT

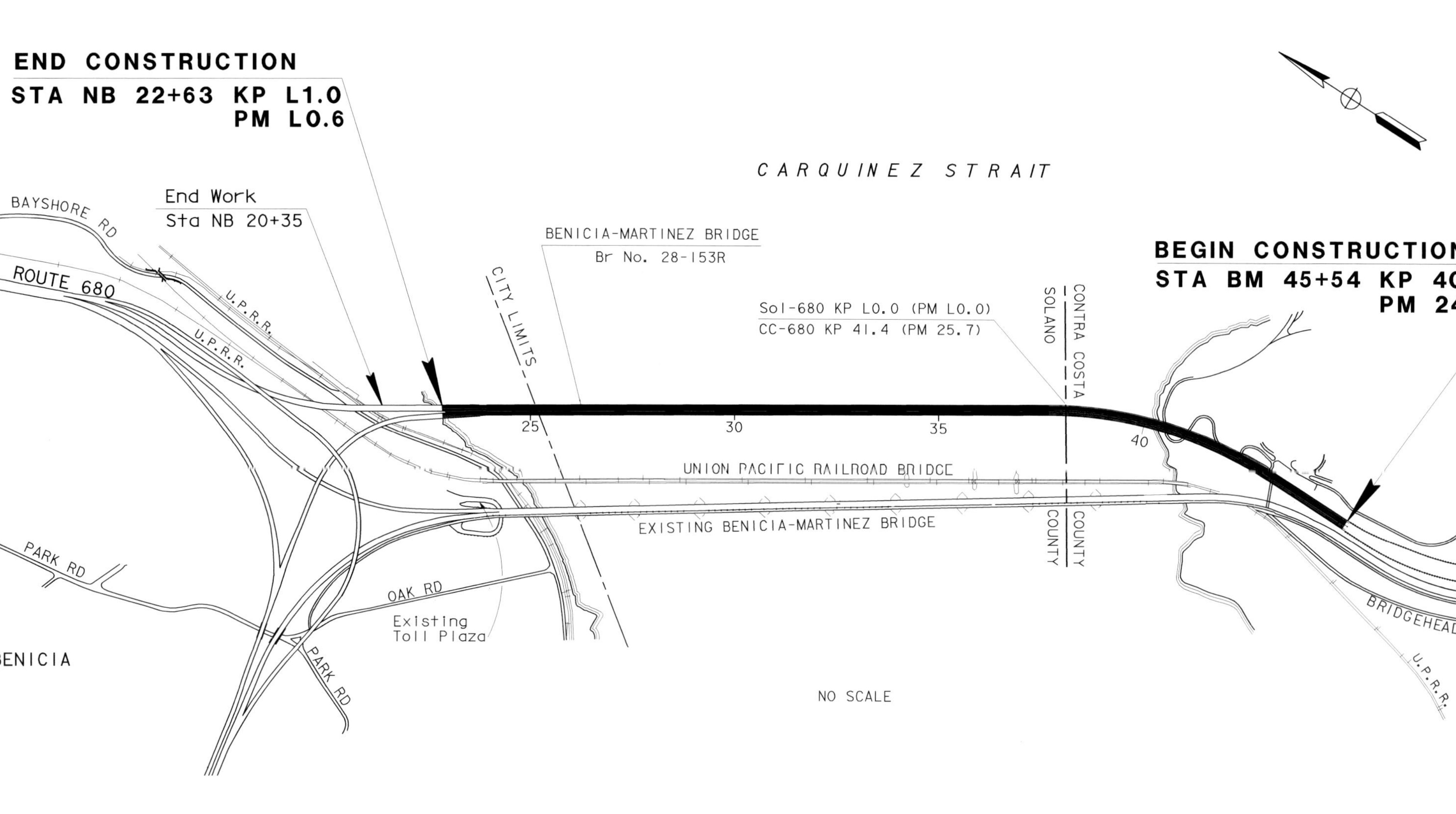

END CONSTRUCTION
STA NB 22+63 KP L1.0
PM L0.6
End Work
Sta NB 20+35
BAYSHORE RD
ROUTE 680
U.P.R.R.
U.P.R.R.
U.P.R.R.
PARK RD
PARK RD
OAK RD
Existing
Toll Plaza
BENICIA
CITY LIMITS
CARQUINEZ STRAIT
BENICIA-MARTINEZ BRIDGE
Br No. 28-153R
Sol-680 KP L0.0 (PM L0.0)
CC-680 KP 41.4 (PM 25.7)
25
30
35
40
UNION PACIFIC RAILROAD BRIDGE
EXISTING BENICIA-MARTINEZ BRIDGE
SOLANO
CONTRA COSTA
COUNTY
COUNTY
NO SCALE
BEGIN CONSTRUCTION
STA BM 45+54 KP 40
PM 24
BRIDGEHEAD
U.P.R.R.

Building the Benicia-Martinez Bridge

TEXT AND PHOTOGRAPHS BY JOHN V. ROBINSON

Carquinez Press

First Printing 2007

ISBN 978-0-9744124-4-3 (hardcover)
      978-0-9744124-5-0 (softcover)
LCCN 2007901320

To order additional copies of this book please contact:
Carquinez Press
P.O. Box 571
Crockett, CA 94525
www.carquinezpress.com

Printed in Singapore

Contents

Foreword

Few people in the construction industry have the opportunity to participate in a project that will stand forever above all others. The Benicia-Martinez Bridge is an example of what can be achieved through teamwork, effort, and determination. For many of us this bridge will be a standard by which all other projects are measured. Over the past five years, hundreds of men and women from several different organizations have worked as a single team to overcome seemingly insurmountable challenges and complete one of the most technologically difficult bridge projects of our time (and the largest of its kind in California). The project team consisted of members from the project owner: Caltrans, designer: T.Y. Lin, and Contractor: Kiewit Pacific Co.

From the construction of pilings up to three hundred feet deep, to the cast in place segmental spans over one-hundred and fifty feet high and up to six hundred feet long, the skill and resolve of the project team was tested at every phase of the project. All obstacles were overcome. We have compiled this illustrated book as a tribute to everyone involved in the successful completion of this project and to commemorate our collective achievement.

Over the years, many people have worked on the Benicia Bridge and moved on in their careers, having benefited from their experience here. In addition to the unique memories that each person will take away from their experience on this project, it is hoped that this book will serve as a reminder of the project team, and be a source of pride for many years to come.

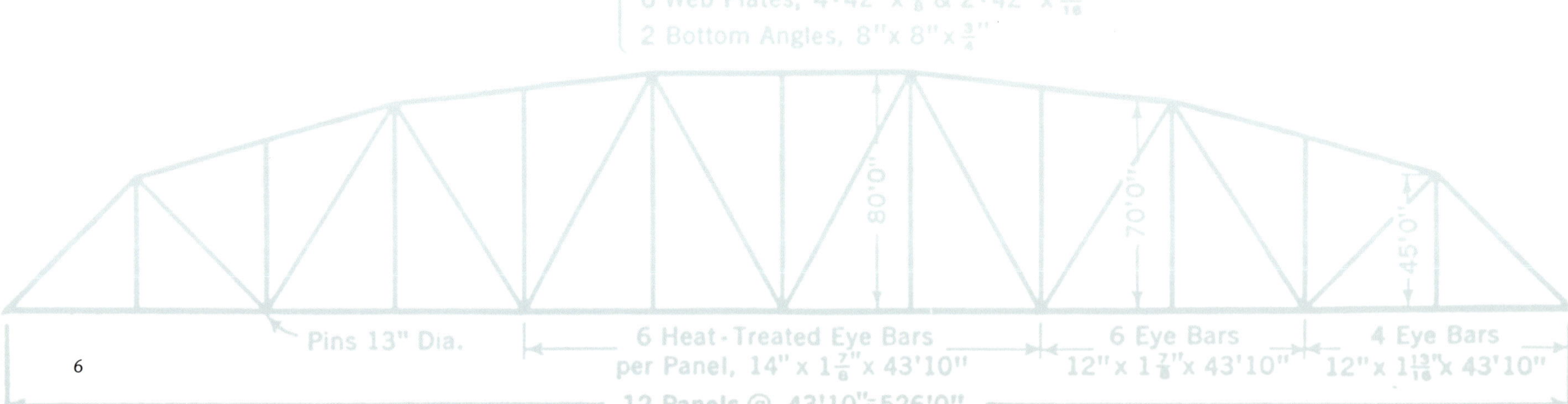

Acknowledgments

I wish I had more space to acknowledge all the people who helped make this book a reality. Over the past 5 years I have met, talked to, and photographed many people. My first Kiewit contact on the Benicia-Martinez Bridge was Ken Riley who, in late 2002, paired me with Howard Mattfield to escort me on my first few visits to the project. I subsequently got side tracked for several months writing *Spanning the Strait: Building the Alfred Zampa Memorial Bridge*. When I returned to the Benicia Bridge Richard Raine was the Project Manager. It was Richard Raine who first put a digital camera in my hands. I want to say "thank you" to Richard for that gift. Since then I have gotten to know several people who have been instrumental in the production of this book: Ron Rattai, Dave Mitchell, Troy Marble, and Rafail Pozin have all played important roles in the production of this book.

Betty Maffei and Lee Taylor of the Contra Costa Historical Society provided research materials about the train ferries, the railroad bridge, and the first highway bridge. Andrea Blachman, of the Martinez Historical Society, also provided research material and historical photographs of the design and construction of the railroad bridge. The Engineering Library at U.C. Berkeley was an invaluable resource for technical information about the train ferries and the train bridge.

Most of all I would like to thank the people who built the bridge. People like: Steve Riley, Tony Dudley, Rick Sanchez, Shawn Frazer, Al Bubion, and many others, encouraged me in my efforts and took the time to explain the work process to me. Were it not for the help of such generous people this book might not have been possible.

John V. Robinson

May 2007

Introduction

Building a bridge is a monumental achievement. I have been fortunate over that past five years to photo-document the construction of the Alfred Zampa Memorial Bridge between Crockett and Vallejo, aspects of the new Bay Bridge in Oakland, the new Tacoma Narrows Bridge in Washington, the deconstruction of the 1927 Carquinez Bridge, and construction of the new Benicia-Martinez Bridge—the subject of this book.

Having so much to do is both exhilarating and daunting. When writing about the old train ferries and first two bridges I consulted publications as varied as: *Engineering News Record*, the *Transactions of the American Society of Civil Engineers*, *Railroad History*, old newspaper articles, and unpublished manuscripts in the collections of local historical societies. The majority of this book is a photo documentary of the new Benicia-Martinez Bridge. Bridges are among our most significant architectural achievements. Bridges unite people and communities. Many books have been published celebrating bridges as architectural monuments. Few books document the construction of our important bridges, and fewer still have documented the contributions of the engineers, piledrivers, iron workers, carpenters, labors, electricians, teamsters, and crane operators, who build our great bridges.

This book is a cooperative effort with Kiewit Pacific's Benicia team. I took the majority of the photographs; Kiewit's Bridge team provided the aerial photographs, taken by Mark DeFeo, and various other miscellaneous photographs. The contractor also provided details about the bridge's construction and other textual elements that I have edited into the text. I have tried to be accurate but I am not an engineer.

Ultimately, this book is a series of judgement calls. I have tried to make the descriptions plain, readable—without being overly technical—and tie them together with the photographs. This book is primarily a photo essay. I have tried to achieve a balance between the technical and aesthetic aspects of the bridge and to allow readers to experience and appreciated the various aspects of the project. I hope you like the result.

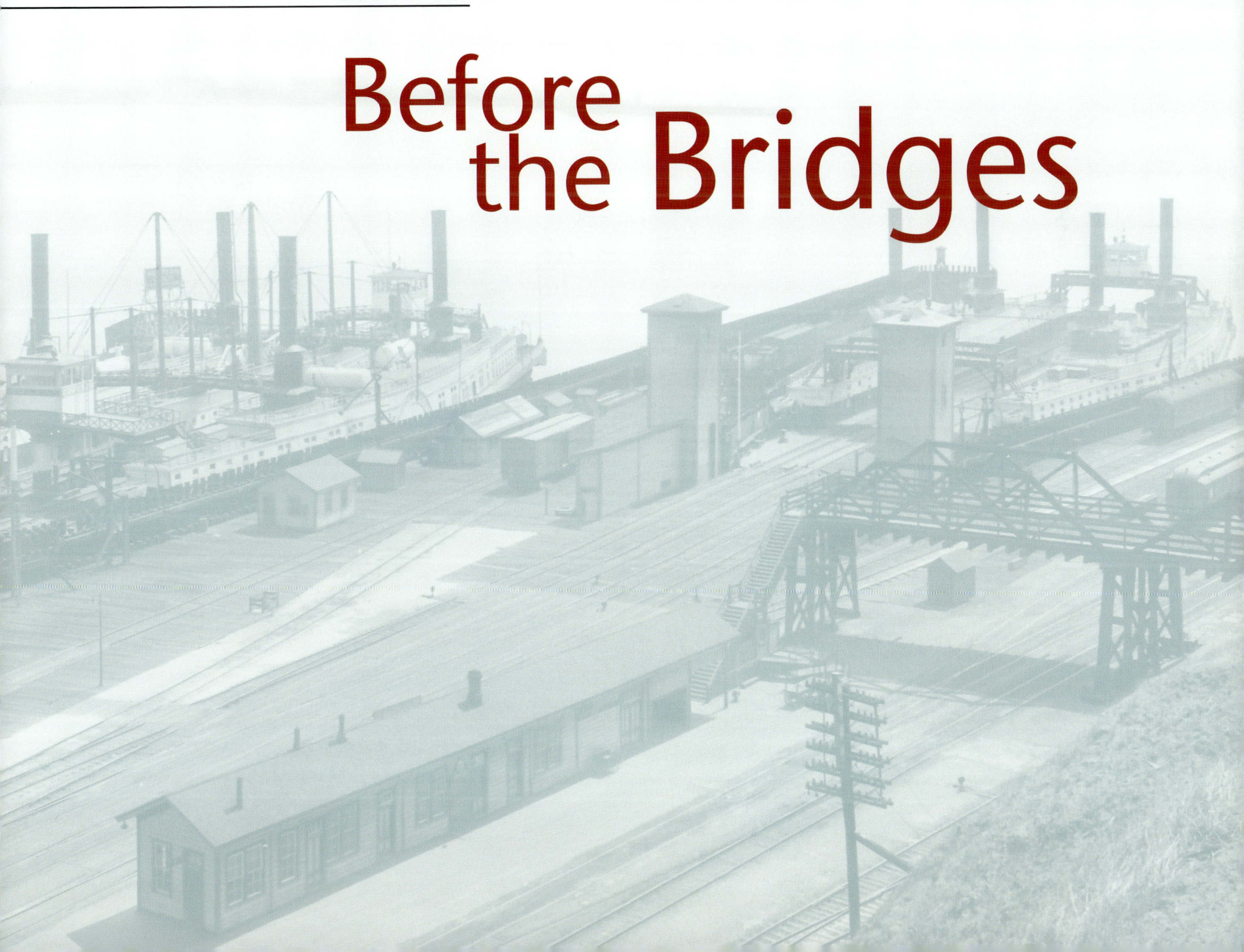

Before the Bridges

The Transcontinental Railroad between Omaha, Nebraska and Sacramento, California was completed in 1869 with the joining of the Central Pacific and the Union Pacific Railroads at Promontory Summit, Utah. By the 1870s the need to get passengers and cargo from Sacramento to San Francisco was growing. This meant going around or going across the vast waterway that makes up the San Francisco Bay. The most direct route was across the narrow waterway of the Carquinez Strait, a six mile stretch of water that starts at Suisun Bay near Benicia and runs west past Port Costa and Crockett into the San Pablo Bay that in turn sweeps south into the San Francisco Bay.

Throughout the years many proposals for bridging the Carquinez Strait were explored by the Central Pacific Railroad. All were rejected as infeasible for a variety of economic and political reasons. Indeed, an interesting book could be written about all the bridges that were proposed but not constructed around the Bay Area in the late nineteenth and early twentieth century.

Finally, in 1878, the Central Pacific Railroad opted to create a train-ferry link across the Carquinez Strait between Solano County and Contra Costa County. To that end, Arthur Brown, Superintendent of Construction for the Central Pacific Railroad, designed the world's largest ferryboat and constructed it in the Central Pacific's West Oakland shipyard. The boat, christened *Solano*, was 425 ft long and 116 ft across the

This early photo of the *Solano* was taken by Carlton Watkins c. 1880.

beam, and displaced 3,500 tons of water. The great ship made its maiden voyage on November 16, 1879 under the command of Captain E. Morton arriving at its newly constructed Benicia Slip with great fanfare at 2:30 P.M. The commander of the Benicia Arsenal welcomed the *Solano* to Benicia by firing a salute of twenty-six cannons.

A mile west of Benicia on the opposite side of the Carquinez Strait a second ferry-slip was constructed at the mouth of Bull Valley. Since it was the Central Pacific's Contra Costa port the little valley was dubbed Port Costa and for the next 50 years the *Solano* ferried freight and passenger trains back and forth from Benicia. The ferry service gave the Central Pacific (later Southern Pacific) Railroad a great advantage over its competitors who moved passengers and freight from Sacramento and Stockton through Tracy to Martinez or over the steep grades of the Altamont Pass as a southern route into Oakland and San Francisco. The *Solano* cut 150 miles and several hours from the trip.

As early as 1901 Southern Pacific Engineers began seriously considering a bridge to replace the aging and overworked *Solano*. A low-level trestle with a 445 ft swing span in the middle of the strait was proposed. The bridge proposal languished for several years before finally being denied by the war department as a navigation hazard. In 1914 the Southern Pacific responded by

The *Solano*, left, and the *Contra Costa*, right, at Port Costa ferry slips c. 1920.

constructing a second massive ferry, the *Contra Costa*, and placing it in service alongside her older sister the *Solano*. For the next 16 years the two ferries worked twenty-four hour a day seven days a week to meet the demands of increased train traffic through the area.

In 1927 a mile-long high-level highway bridge was opened between Crockett and Vallejo at the western end of the Carquinez Strait. The mighty Carquinez had been spanned just as the two great ferries were reaching the end of the working lives. Indeed, the *Solano* had already outlived its predicted service life by more than 30 years—kept in service by necessity. By the late 1920s the trains were longer, the engines were getting much heavier, and the ferries that once expedited travel were now creating a bottleneck as the two boats struggled to meet the workload. To meet the increasing demand, the *Solano* would need to be replaced and a third ferry put into service. Since there was no room for a third ferry in the narrow confines of the Carquinez Strait a bridge was the only solution.

The *Contra Costa* lumbers into the Port Costa ferry slip c. 1925. Notice the list of the ferry due to the weight of the larger locomotives.

CHAPTER TWO

Bridging the Suisun Bay

After years of false starts, surveying various sites, and weighing their options, Southern Pacific Railroad chose a point on the mouth of the Suisun Bay at eastern edge of the Carquinez Strait. The new bridge would connect Solano and Contra Costa County between Army Point in Benicia and Bulls Head point in Martinez. The chosen site had solid rock footings for the abutments and heavy piers and was close to the existing train lines.   The design work began in 1927 and a groundbreaking ceremony took place on May 3, 1929.

The new bridge would be the largest train bridge west of the Mississippi and one of the last large bridges built by the railroad. The contractor Siems, Helmers, and Schaffmer of St. Paul, Minnesota built the substructure which consisted of two abutments, 22 viaduct pedestal piers, and ten large water piers to support the truss spans.

There were three preferred methods for building bridge piers at the time of the first bridge's construction. In shallow water, cofferdams could be built, the water pumped out—creating a hole in the water—and the pier is built in the empty pit. This is how the shallow water piers 10 and 11 were constructed. Another solution is the pneumatic caisson in which a hollow chamber is sunk to the bottom and compressed air is used to keep the water out. Men working in the chamber excavate

In this picture from April of 1929 we see the work continuing on the pedestal piers. Pier 11 is complete in the center, and the steel cylinder for the sand island is in place at Pier 12. Courtesy Martinez Historical Society.

out the mud and as the pier is built on top of the caisson it slowly settles to bedrock. This method is dangerous and men cannot safely work at depths of over 100 feet without suffering the effects of "caisson's disease" first identified on the Brooklyn Bridge and what is known today by divers as "the bends." It is simply the adverse effects of coming out of a pressurized environment without sufficient time to decompress. The swift current, thick layer of mud, and deep-water made pneumatic caissons impractical. The third, and preferred, method is the open dredge caisson. Here the caisson is placed on the bottom with the top protruding from the surface. As material is dredged out of the caisson the pier is built up and the weight of the caisson and pier allows the structure to sink through the mud to bedrock.

This third method was the best option for the new train bridge. But the swift current made placing the caissons a difficult proposition. A few years earlier engineers at the Carquinez Bridge at the western edge of the strait had difficulty placing their center pier caisson. Aware of this fact, engineers at the Suisun Bay Bridge employed a "sand island" method of building the piers. This technique had never been used before in such deep water. Workers built a large steel cylinder, 88 ft in diameter, lowering it down until it reached from just above the surface to the mud line. Initially the cylinder remained

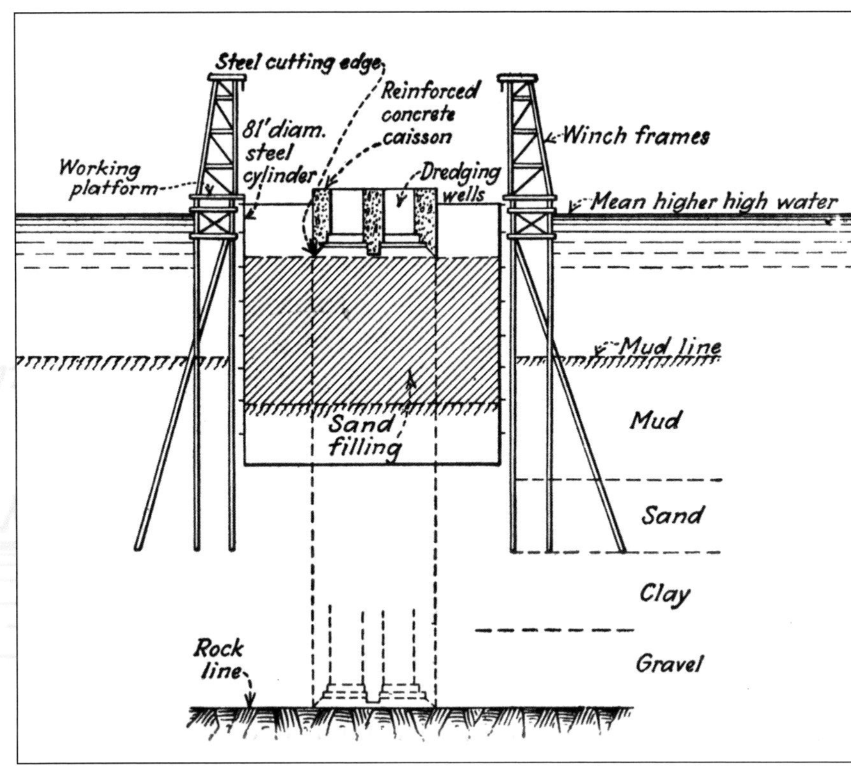

A drawing that illustrates the sand island method of sinking the caissons.

filled with water. When the cylinder was complete it was filled with sand and gravel to displace the water. This created an artificial island on which workers could build the caisson and pier.

Workers began constructing the 40 x 60 ft steel cutting-edge and six concrete reinforced dredge wells on top of the sand island. As more concrete and steel were added to the caisson the sand was removed from the dredge wells and the caisson sank under its own weight through the sand island, through the mud, and down to bedrock. Periodically, deep-water diver Bill Reed was sent down to inspect the cutting edge of the caisson, which he reported back to the surface via a phone-line in his helmet. Reed also brought back samples of the rock for geologists to examine. It is believed this was the first time a deep-water diver was used to inspect piers at such great depths. Bill Reed went on to great fame when he made even deeper and more dangerous dives on the Bay Bridge project in the mid 1930s. Once the caisson was in place the dredge wells were filled with concrete and the pier was built on top of the caisson. Once construction of the pier was well underway, pieces of the steel cylinder above the mud line were unbolted, removed, and used to construct the next sand island. In this manner, the deep-water piers were constructed safely and efficiently.

A diver and his tenders preparing for a dive.

Diver Bill Reed, second from the right, confers with engineers after one of his many dives.

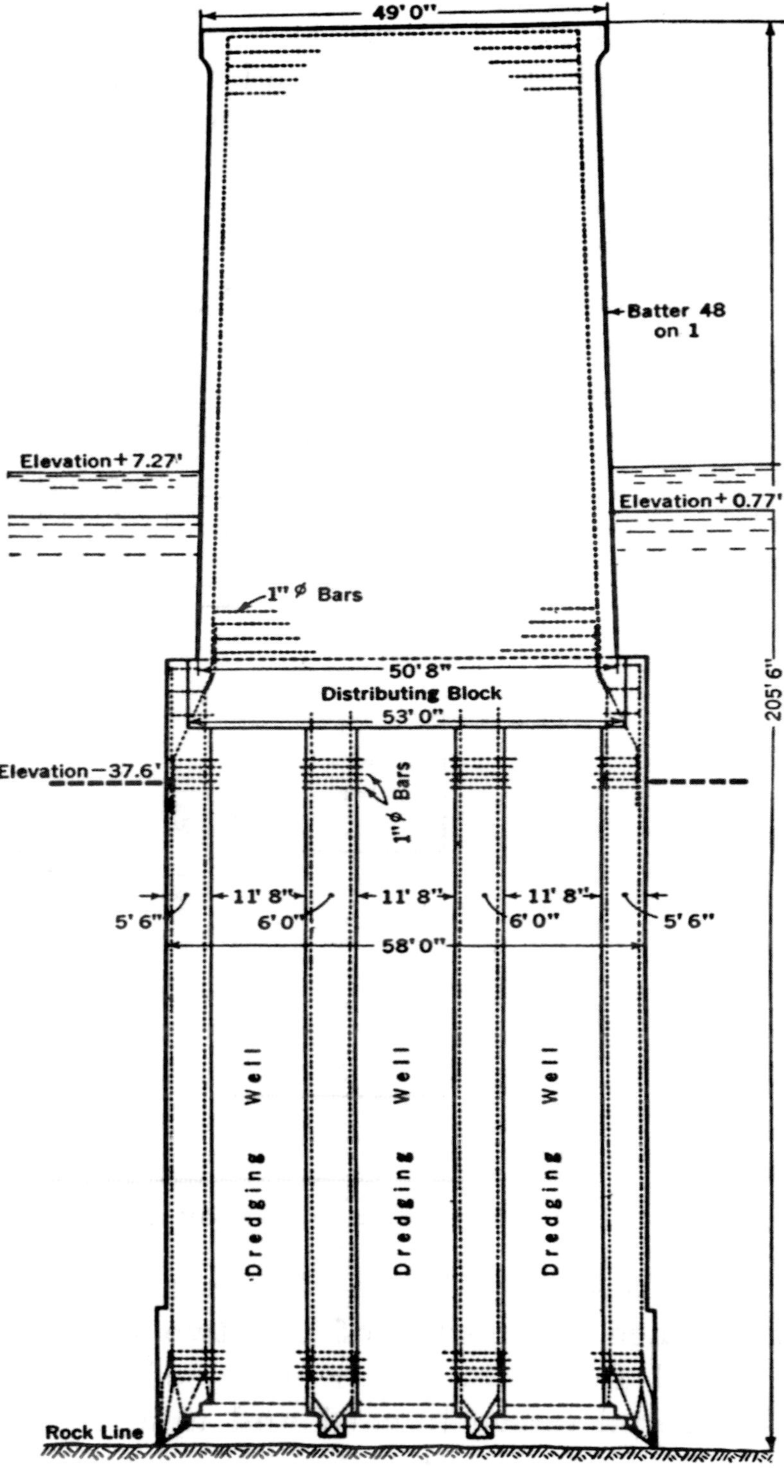

This drawing shows the scope of Pier 13 from bedrock to pier top.

This photograph from February 1930 shows the forms of Pier 12 set to final height. Courtesy Martinez Historical Society.

The towers that raise the 328 ft lift span are supported on Piers 12 and 13. Pier 13 at the north end of the lift span is the deepest at 205 ft from bedrock to pier-top. Its base area is 40 by 60 ft and the pier weighs in at 19,000 tons. The steel structure of the bridge consists of seven identical 526 ft Warren truss spans, one 328 ft vertical lift span, a 560 ft viaduct approach, a 266 ft deck-truss span on the south shore, and a 506 ft deck-truss span at the Benicia shore. Working north from the Martinez shore American Bridge began setting structural steel as soon as the first pedestals and piers were complete.

In November of 1929 the first girder was set in place on the Martinez shore. The completed Piers 11 and 12 are visible in the distance. Courtesy Martinez Historical Society.

Another innovative construction method used to build the bridge was using the 506 ft north deck-truss viaduct as a temporary false work to construct the seven Warren truss spans. The viaduct was constructed on piers near the Martinez shore, floated by barge into place between the piers and lowered onto temporary supports. This created a stable erection platform from which to construct the trusses. When the truss could support itself the erection span was floated to the next set of piers and the process repeated. Since all the trusses were the same and the piers were same distance apart work progressed quickly. When it came time to build the 328 ft vertical lift span the erection span was floated back onto its construction pier, a 300 ft portion removed and floated between Piers 12 and 13 to support construction of the lift span. When all the Warren truss spans were completed the erection span was put into place between Pier 19 and the north abutment.

Floating the erection span into place between Piers 11 and 12. Courtesy Martinez Historical Society

The erection span is in place between Piers 11 and 12 and work on the superstructure is progressing well. Courtesy Martinez Historical Society.

When the lift span was complete the erection span was reassembled and followed the construction progress north until it was finally put in place between Pier 19 and the north abutment. Courtesy Martinez Historical Society.

Using so many innovative construction techniques allowed the 5,600 ft, $12,000,000 bridge to open only 18 months after start of construction. The first trains rolled across the new bridge on October 15, 1930. Courtesy Martinez Historical Society.

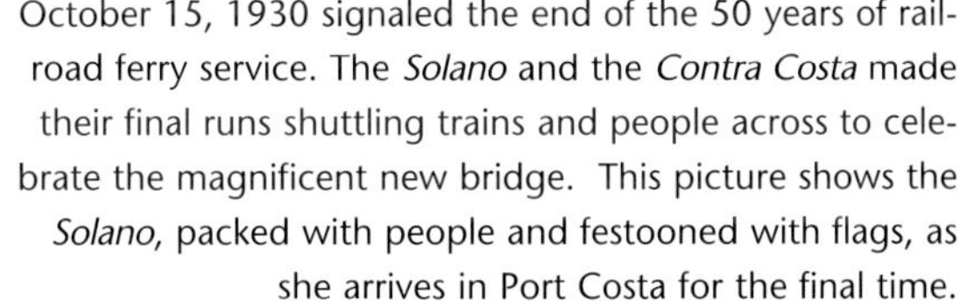

October 15, 1930 signaled the end of the 50 years of railroad ferry service. The *Solano* and the *Contra Costa* made their final runs shuttling trains and people across to celebrate the magnificent new bridge. This picture shows the *Solano*, packed with people and festooned with flags, as she arrives in Port Costa for the final time.

The official bridge opening was November 1, 1930. In this photograph dignitaries and local celebrants greet the historic C.P. Huntington No.1 as the first official train across the bridge, followed by a modern locomotive. The C.P. Huntington now sits proudly in the California State Railroad Museum and is depicted on the Museum's logo. It is the oldest locomotive in the museum's collection and only surviving standard gauge 4-2-4 Locomotive in the United States. The master of ceremonies was James Rolph, Mayor of San Francisco and soon to be Governor of California. James Rolph died in office on June 2, 1934. The San Francisco- Oakland Bay Bridge was named the "James 'Sunny Jim' Rolph Bridge" in his honor when it opened in 1936.

The overall celebration included an air circus to coincide with the opening of Clyde's new Airport. At noon the Queen of Progress, Miss Edna Harrison of Martinez, christened the new span with a bottle of sparkling water. It was prohibition era and no champagne was available. Pictured here are the Bridge Queen Candidates, from Martinez are: Molly Cabral, Judy Bettencourt, Edna Harrison, Rose Lucido, Geniveva Machado; from Pittsburg are: Thelma Smith and Dodo Perry; representing Walnut Creek is Ruth Bunce, from Crockett is Alvera Panuzzi, and from Concord is Mildred Miller.

By early 1961 a small armada of tug boats, barges, and barge cranes was at work building the first highway bridge between Benicia and Martinez.

By May of 1961 the first road-deck truss is reaching out from the shoreline towards the first water pier. Notice steel falsework supporting the truss and the safety nets used to protect the workers.

By the late 1950s the state recognized the need to expand the highway system in the Bay Area. In 1958 a second Carquinez Bridge was opened between Crockett and Vallejo. The next project was an auto bridge between Martinez and Benicia. Yuba Consolidated Industries built both the substructure and the super-structure of the 14.2 million-dollar project.

The 6,215 ft bridge consist of ten concrete piers which support seven 528 ft steel-truss spans, two 429 ft spans, and one 330 ft span with steel girder approaches at each end. The bridge has 138 ft vertical clearance over the shipping channel. When the bridge opened in September of 1962 the road deck was 62 ft wide with four lanes of traffic. In 1970 the bridge was christened the George Miller Jr. Memorial Bridge in honor of State Senator Miller who died in office in 1969.

This shot from 1961 shows one of the truss panels being set in place atop one of the concrete piers. The safety net is already partially attached and will be pulled taunt when the next panel is set in place allowing iron workers to safely set the stringer beams between the two truss panels.

In this detail shot from 1961 we see ironworkers setting the steel between the truss panels, we get a better look at the safety net and the temporary falsework supporting the deck truss. The safety nets allowed workers to work without fear of falling. The nets also prevented tools and materials from falling into the water.

This easterly view from January 1962 shows the deck trusses extending halfway out into the strait. The stiff leg derrick "Judy Ann" mounted on a barge is being used to set the steel.

The opening of the new bridge put an end to the oldest ferry service on the bay. Martinez had a ferry service to Benicia, then the state capital, starting in 1854 when the steam-ferry *Carquinez* was put into service between the two towns. In 1860 the *Carquinez* transported "Pony Express" riders from Benicia to Martinez on their way to Oakland and San Francisco. The service ran intermittently over the next hundred years and many different ferries were used. In 1953 the State bought the service and commissioned a ferry, also called the *Carquinez*, to transport people and autos between Martinez and Benicia. When the bridge opened the ferry was sold to the state of Florida.

This ticket for the last ferryboat ride was issued to commemorate the termination of one of the oldest ferry systems in the Bay Area.

In this c. 1970 photo Congressman George Miller is seen at the unveiling of the sign renaming the bridge the George Miller Jr. Memorial Bridge in honor of his father, State Senator Miller, who died in office in 1969 and was instrumental in getting the highway bridge built.

This December 2004 photograph shows the gentle arch of the mile-long truss-deck bridge. From 1989 to 1991 Kiewit expanded the road deck to accommodate six lanes of traffic. When the new bridge opens the old bridge will be reconfigured to carry four lanes of southbound traffic with a bicycle lane and a pedestrian walkway.

CHAPTER THREE

NEW BRIDGE:
Site Preparations
and Early Construction

The new Benicia-Martinez Bridge consists of four contracts: the new I-680 approach on the Martinez side, the new toll plaza on the Martinez shore, the I-680 / I-780 interchange near Benicia, and the largest of the four, Kiewit Pacific's contract to build the 7,500 ft bridge.

The contract for constructing the 7,500 ft bridge between Contra Costa County and Solano County was awarded to Kiewit, a joint venture of four Kiewit districts: Northern California, Pacific Structures, General Construction, and Kiewit Engineering Co. The owner's team consisted of California Department of Transportation (Caltrans), with engineering management/support from the designer of the bridge, TYLin.

Notice to proceed was given on October 25, 2001. Many weeks of site preparation were required before construction could begin. Access roads to the shoreline were constructed and tons of material and tools were brought to the site. A staging area was setup in Vallejo at the western end of the Carquinez Strait. The Vallejo site was used to store and inventory the thousands of tons of steel falsework needed during the construction of the bridge, the tower cranes, and the traveling form system that would be used to build the concrete segments. The Vallejo yard was also used by Kiewit's subcontractor Regional Steel to build all the rebar pile-cages and pier columns. The Vallejo staging area allowed necessary

This photo from September 2003 shows some completed rebar pile cages ready to be barged to the bridge site. (Photo by: heliphotos@mchsi.com)

materials to be close at hand without cluttering the narrow bridge site. Necessary material could be quickly barged to the bridge site as needed from the Vallejo yard.

As the bridge neared completion the Vallejo yard became a storage site for demobilized equipment and materials.

The Martinez side is hemmed in by the existing bridges on one side and a chemical plant on the other side. Marshy terrain, steep bluffs, and railroad activity made the Benicia shore equally unsuitable for large scale storage of equipment.

Before any real construction could begin access roads had to be cut to the site and trestles had to be built on both shores through the tule marshes to gain access to water deep enough to accommodate crew boats, tug boats, and barges.

Across from the Vallejo yard, at Mare Island, another work area was setup to construct the Stage One pier-footings that sit over the foundation piles and support the massive columns and pier tables.

By September of 2003 the Vallejo yard was filling up with steel falsework and other materials needed to build the bridge. (Photo by: heliphotos@mchsi.com)

This photograph from September 2003 shows one of the twelve 1,800-ton pier footings that were constructed at the Mare Island site. (Photo by: heliphotos@mchsi.com)

This view looking north shows access to the Martinez shore. Piers 1 through 4 and 1,000 ft of bridge were constructed through this narrow corridor.

The Benicia shore presented different access problems. Steep bluffs and railroad tracks crowd the shoreline. A thick marsh of tule-reeds stretch about hundred feet out from the shore. A large trestle was constructed from the Benicia shore to gain access to open water and support the falsework for the cast-in-place portions of the bridge at Piers 16 and 17.

On the Martinez shore a similar trestle was constructed to reach out to Pier 6 as well as provide access for crew boats and barges.

On the shore above the trestle near Pier 5 a large concrete batch plant was assembled to provide the special lightweight concrete mix needed to construct the bridge segments.

Mission Control

While the early site preparations were underway a team of people was assembled to resolve construction conflicts, generate shop drawings, and document contract changes. The team was called "Mission Control." The goal of Mission Control was to monitor each phase of the project and anticipate, eliminate, or minimize field delays. The team was composed of construction and design staff. Representatives from Caltrans, TY Lin International/CH2M Hill JV (Owner's Engineer), Kiewit (Contractor), Parsons Transportation Group (Contractor's Engineer), and Regional Steel Corporation (Contractor's reinforcing steel subcontractor).

In 2002, during the early stages of detailing shop drawings for Pier Table 5, it became clear that the large amount of reinforcing steel and post-tensioning cables would conflict with some of the embedded items needed for construction purposes. The structure of the

By July of 2002 the footing for Pier 5 was being poured. Pier 5 was the first Pier constructed and was used, via a mock-up, as a model for the construction of the other Piers.

bridge does not just support itself, but has to support falsework, concrete forms, work platforms, and other equipment used to construct the bridge. Sometimes the items used in the construction of the bridge create points of congestion in, and on, the bridge that need to be resolved so work can proceed safely and efficiently.

There are various ways to address these issues. One is the use of 3-D computer models to help visualize the problems and plot various solutions. Another way to work the problem, and the method chosen for Pier 5, was to build a full-scale mock-up of one-half of Pier 5 and allow the team to work through the conflicts in a hands-on setting. The pier table mock-up was time consuming, but beneficial. Once the conflicts were solved with Pier 5 it was easier anticipate and correct problems that would arise in the construction of the other piers.

Mission Control was in place for 18 months and during that time over 400 issues were successfully resolved.

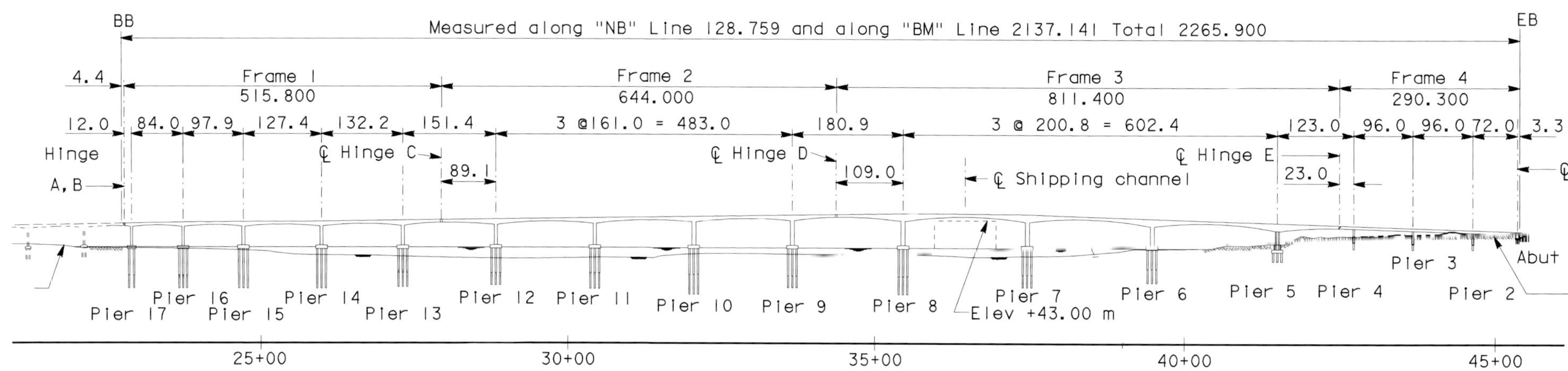

DEVELOPED ELEVATION

1:5000

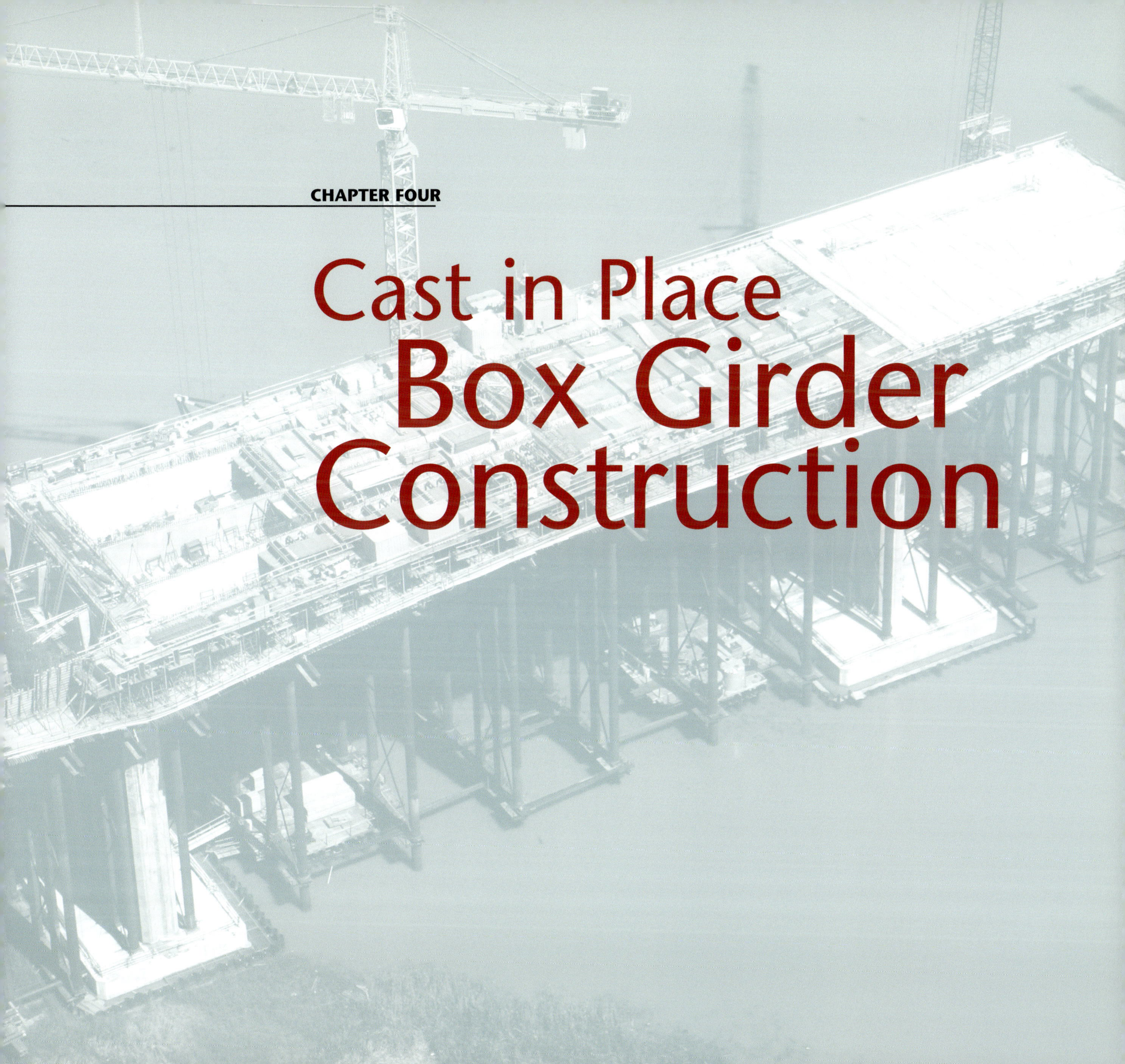

Cast in Place
Box Girder
Construction

In addition to the main span of the bridge the contract included a 1,000 ft viaduct approach starting in Martinez at Abutment 1 to the first hinge joint (Hinge E) north of Pier 4. Spans 1 and 2 were constructed on wooden falsework while the section from Pier 3 to Hinge E was constructed using steel falsework. The entire structure from Pier 1 through Hinge E was known collectively as "Frame 4."

By December 2002 the rebar for the bottom deck and walls of Frame 4 was complete through Pier 2. The continuity post-tensioning ducts are also in place in the stem walls

This photo from December 2002 shows the wooden forms supporting Piers 1 and 2 of Frame 4. The tower crane is set on rails parallel to Frame 4 so it can freely travel up and down the frame as needed. The narrowness of the space made the use of a traditional crawler crane impractical.

Iron workers consult drawings while installing rebar at Frame 4 above Pier 2.

Installation of steel falsework at Pier 3. As the piers got higher, steel replaced wood as falsework material. Steel is stronger than wood, takes up less space, and creates less congestion under the structure.

Concrete pour on the bottom deck of Frame 4 near Pier 2 is underway in March of 2003.

Looking north in this aerial view from September 2003 we see form and concrete work continuing on Frame 4. The column of Pier 5 is complete and work is progressing on some of the other piers. The first column is under construction at Pier 9. (Photo by: heliphotos@mchsi.com)

The north section (Frame 1) in Benicia is over the water and required pile up to 120 ft deep to support the pipe falsework supports. The bridge widens at the north end where it ties into two separate bridges from a separate Caltrans project. The easterly connecting bridge is I-680 north toward Sacramento and the westerly connecting bridge is I-780 toward Vallejo. The complexity of the Benicia box construction is notable because of the combination of many design elements and transitions.

By November of 2005 a large and complex steel falsework was in place to support the bridge where it connects to the shore and ties into the I-680/I-780 interchange.

A view from inside the form structure atop the cast-in-place box girder section of the bridge.

Looking north we see the complex design of Frame 1 where it ties into the I-680 and I-780 interchange at the Benicia shore.

CHAPTER FIVE

Concrete

Batch Plant

The three main components of the bridge are reinforcing steel, post-tensioning cable, and concrete. The concrete was mixed on site at a large batch plant near Pier 5 on the Martinez shore.

The aggregate and sand used in the various concrete mixes were stored on site in large silos.

A conveyer system fed the material to the batch-plant where the concrete was mixed and loaded onto trucks.

Kiewit employee monitors the concrete mix from the control room of the batch plant.

Various normal weight concrete mixes were used to construct the foundations, piers, columns, and pier tables.

Constructing the segments between the pier tables presented new challenges to the team. To meet the design requirements for strength and weight a mix was developed that is 20% lighter than normal structural concrete. This lightweight concrete consisted of cement mixed with coarse aggregate from North Carolina and sand from Canada. But the lightweight mix came with a tradeoff—high internal curing temperatures.

Concrete Cooling

Concrete heats as it cures. This is a well-understood process that engineers can use to their advantage. For example, by measuring the internal temperature of a concrete structure engineers can determine when it is safe to remove forms and apply a load to the structure.

The curing temperatures in the segmental portion of the bridge were high enough that thermal control plans were needed to limit overheating of curing concrete and prevent cracking. One solution was to embed cooling tubes in the concrete of the segments. Water from the bay was pumped into the cooling tubes to dissipate heat as it flowed through the segments back into the bay.

This view looking north from inside Pier 9 shows elements of the cooling manifold at the bottom and the right of the frame during a concrete pour from November of 2005. When the segments were complete the embedded cooling tubes were filled with grout.

Another solution to the overheating problem was to cool the concrete itself before it was poured. A portion of water in the concrete mix was replaced with ice to cool down the concrete. Blocks of ice, weighing approximately 300 lbs, were shaved and added to the concrete while it was being mixed. In cooler months, 30% of the water in the concrete mix was replaced with ice to cool it to 65 degrees when loaded into the concrete trucks. The addition of ice into the concrete mix is the most cost-effective method to cool it.

A third solution utilized liquid nitrogen to control the temperature of the concrete mix. Concrete trucks would drive from the batch plant to a cooling station that looked like an oversized automated car wash. At the cooling station liquid nitrogen was sprayed into the concrete mixer to keep the load cool until it could be delivered to its destination.

During the summer months, liquid nitrogen was used in conjunction with ice since the thermal control plan requires cooler temperatures than ice alone could provide. On the hottest days, the concrete was cooled to as low as 35 degrees and peaked at temperatures of 160 degrees as it cured.

The concrete trucks delivered their load by barge. The trucks were loaded onto barges at the end of the south trestle and brought by tugboats to a pump-barge, with a 58-meter pump truck, a conveyor belt, and a concrete re-mixer. When the trucks arrived at the pour site the pump-truck pumped the concrete up to the bridge.

A truck in the nitrogen cooling station. The man in the center checks the drum's temperature with a hand held device.

Workers monitor the concrete's temperature as a truck passes through the cooling station.

Concrete trucks were brought to the piers by barge. Once there, another barge with a pump-truck received the load and pumped it up to the segments.

Empty trucks are barged back to the south trestle to receive another load of concrete.

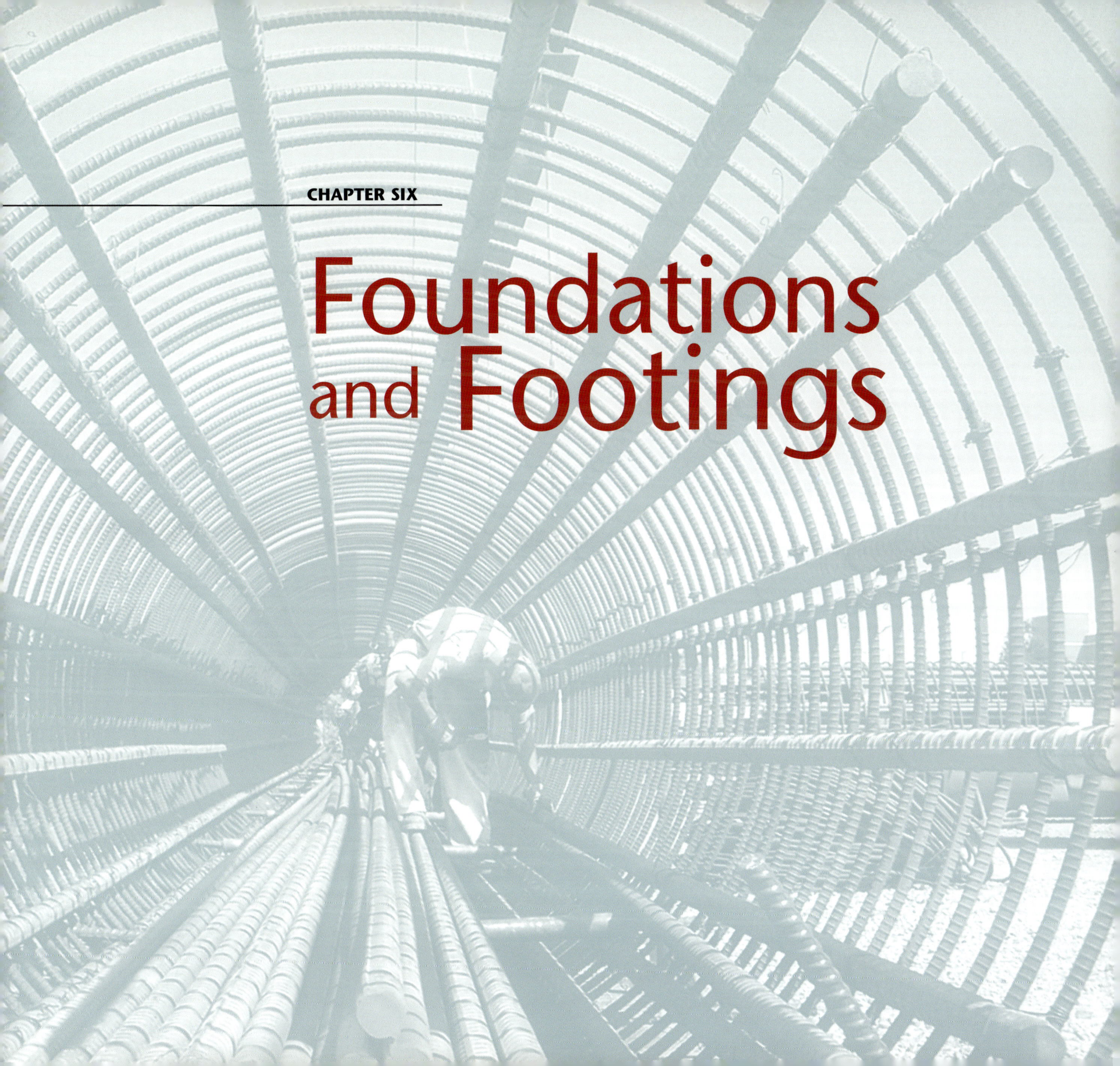

Foundations and Footings

There are a total of 17 piers at the Benicia-Martinez Bridge. Piers 1 through 5 were land-based, while piers 6 through 17 were constructed in the water. Each marine pier was founded on either eight or nine steel piles—depending on the location of the pier and the load it would carry.   The foundations for the marine piers were constructed in four main steps: driving the 8.5 ft diameter steel casing through the mud to the bedrock, drilling the 7.2 ft diameter rock-socket into the bedrock, and placement of rebar cages in the rock-sockets and steel piles, and pouring the concrete.

Casings and Rock Sockets

The installation of the steel casings was achieved with the aid of a 600,000 pound, 70 x 80 ft driving template. The template acted as an island that did not fluctuate with the tide or current and provided the necessary stability for the casing. Once the template was installed, the casings were lofted with a derrick barge and swung into the pockets on the template. The casings then sank under their own weight into the mud. After the casings were stabbed and plumbed, a 135-ton hydraulic impact hammer, with 400,000 foot-pounds of energy, drove them to tip.

In April of 2002 it was discovered that the concussion from the pile-hammer was killing some of the salmon, steelhead, sturgeon, and striped bass that make the

Here we see a complete eight-pile spread and the template-held in place in place by four temporary "spud piles" at the corners.

Second largest pile-hammer in the world drives casing to bedrock. A bubble-curtain of air protected fish from the massive concussion generated by the hammer.

Carquinez Strait their home. Kiewit, working with Caltrans and an outside design firm, devised a simple, but ingenious, solution to counteract the problem. A frame of perforated pipes, called a "bubble-curtain" was constructed to surround the piles. When compressed air was pumped into the pipes a "Jacuzzi-like" curtain of bubbles formed around the pile and dampened the concussion from the pile hammer.

The bottom-mud was 70 to 120 ft deep and was comprised of silts, sands and clays. Below this overburden was the bedrock, a sedimentary rock formation comprised of shale and sandstone. The pier-shafts penetrated the bedrock anywhere from 80 to 130 ft deep.

One section of the bubble-curtain sits on a barge.

At some locations, the hard rock caused casings to deform during piledriving. These had to be repaired before the rotator work could begin.

The contract called for a pile-top reverse circulation method for drilling the rock sockets. In reverse circula-

Two crewmembers in spider lift guiding out a piece of damaged casing. When a case got bent the damaged portion was cut out and repairs made. A section of damaged casing is seen on the trestle near Pier 6.

tion a thick polymer-slurry is pumped into the hole as
the auger penetrates the rock. As the auger-bit grinds
deeper into the bedrock, the sediment laden slurry is
pumped up from the bottom through the drill shaft,
sluiced through several chambers on a barge to clear
the sediment and pumped back into the top of the
hole. The slurry pressure helps support the hole as the
auger pushes deeper into the rock. This technique was
successful at Pier 9—the first location of rock scoket
drilling.

Once drilling started on the second pier location
unstable bedrock was encountered that made reverse
circulation drilling impossible—the slurry pressure
could not keep the shafts from collapsing.

To combat this issue, Caltrans directed Kiewit to change
to the rotator method of drilling the rock sockets.
Unlike the reverse circulation method, which is drilled
through an uncased hole, the rotator method is drilled
through a steel-cased hole. In this method the steel case
itself is rotated and bores itself into the bedrock.

As the steel casing descends into the earth a new section of casing is attached at the top of the rotator and

Rotator machine at work on Pier 14.

Here we see a rotator machine, and blue crawler
crane inserting the clamshell attachment used to
excavate rock at the bottom of the casing.

Two sections of rotator used to construct the rock sockets on a barge. The section to the right, called the "starter" casing has teeth that are used to bore into the bedrock.

the process continues until a satisfactory depth is reached—some of the casing reached 300 ft into the bedrock. The purpose of the rock sockets is to root the foundation of bridge in the bedrock.

Once the rock sockets were drilled, the rebar cages were inserted into the casing and concrete was poured. As the concrete went in, the steel jacket was slowly removed and disassembled to be used in the next hole.

A frame constructed and mounted on a barge was used to tip the rebar cages and columns upright before the cranes lift and set them in place. Here a rebar pile-cage is ready for insertion into the pile casing.

To perform the rotator drilling work, Kiewit sought the services of Malcolm Drilling Co. of San Francisco. Drilling rock sockets via the rotator method is common on land, but not on water. In order to counteract the approximately 5 million foot-pounds of torsion created by the rotator, an 1,800 ton platform was designed to sit on the pile casings that were already driven to bedrock. The rotator platforms provided a stable perch for Malcom's heavy drilling equipment. Two rotator platforms were built for deep-water piers and moved from pier to pier as needed by the catamaran barge.

When the task was finally complete: 27 million pounds of steel casing had been inserted; 22,000 linear feet of shafts drilled; 33,000 cubic yards of spoil removed from the holes; 12,000 tons of rebar cages were installed into the shafts, and 41,000 cubic yards of concrete filled the pier shafts to provide a stable foundation for the bridge superstructure.

Deep Water Footings

The deep-water footings, Piers 7 through 15, were constructed in two stages. The first stage was pre-cast on a barge at the Mare Island yard.   Each footing is 69 ft x 64 ft x 4.5 ft deep. After the stage 1 rebar, concrete, and post tensioning work was completed, the stage 2 watertight forms were installed on the pre-cast footing. The footing was then towed out to the bridge and rigged up to the catamaran barge.

Catamaran barge lowers stage 1 footing into place at Pier 15.

The catamaran barge then lifted up the 1,800-ton stage 1 footing, fleeted over the top of the 8 or 9-piles cluster, and then lowered the footing onto the pile cluster.

Once the footing was lowered onto the piles the space between the pile and footing was sealed, the water pumped out, and grout was placed in the annulus space. Once cured the piles and grout supported the

footing. With stage 1 securely in place, the stage 2 rebar, the column template, and a portion of the column rebar was placed.

Finally, the stage 2 concrete was poured to a depth of 13 ft. When the concrete cured the forms were stripped and the footing was post-tensioned. The nine deep-water footings contained 22,680 cubic yards of con-

This view from July of 2005 shows a Regional Steel iron worker amidst the stage 2 rebar at Pier 14.

A forest of purple rebar sticks up from one of the stage 1 footings.

crete, 5,310 tons of rebar and 13.5 miles of 4" and 6" diameter post-tensioning duct.

Shallow Water Footings

Pier 6, near the Martinez shore, and Piers 16 and 17, near the Benicia shore, were shallow water footings. Both stage 1 and stage 2 of the shallow water footings were constructed from the work trestles. The stage 1 footing was cast on falsework that set directly over the pile grouping. Once the stage 1 forms, rebar, concrete and post tensioning work was completed the stage 2 forms were installed and a lowering system was connected to the footing.

The falsework was then stripped out below the footing and above the pile grouping. The 1,800 ton footing was then lowered into position using four 400 ton hydraulic rams. Once stage 1 was in position, the mud and silt were removed from the space between the piles and the footing. Water was then pumped from the footing and grout was placed in the annulus between the piles and the footing. With the grout in place the footing could

support itself on the piles and the lowering system was removed.

Stage 2 of the shallow water footings was constructed in the same manner as the deep-water footings: Rebar and post-tensioning duct were installed and 13 ft of concrete poured. When the concrete cured the forms were stripped, post-tensioning strands were inserted into the ducts and the footing was post-tensioned.

The three shallow-water footings contained 7,500 cubic yards of concrete, 1,770 tons of rebar and 3.75 miles of post-tensioning duct.

Here we see the Pier 6 footing on falsework above the piles.

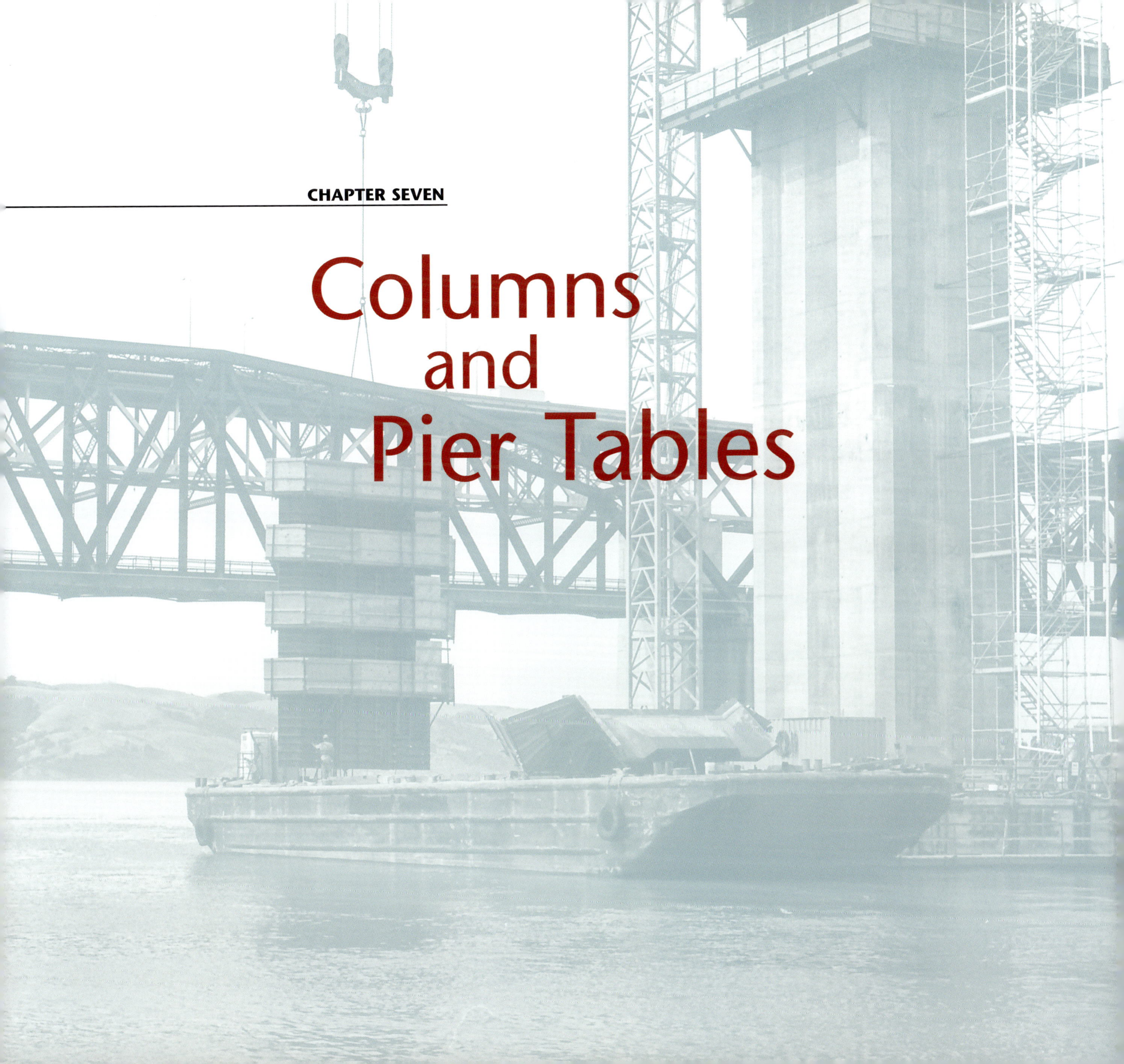

Columns and Pier Tables

Columns

The new Benicia-Martinez Bridge is supported on 16 column lines. The land columns are solid, while the water columns are hollow with an interior stair tower that runs from the pier footing to the pier table. The hollow columns average 120 ft tall and max out at 136 ft above the shipping channel. The large columns contain as much as 250 tons of rebar and 1,100 cubic yards of concrete. In total, the 16 columns that support the new bridge contain over 2,500 tons of rebar and 15,000 cubic yards of concrete.

Lifting a rebar cage into place at the Pier 6 footing.

Setting the last of four rebar cages that will make up the first stage of the Pier 6 column.

Once the four corners of the column were set, the iron workers began placing the horizontal steel that would tie it all together and make the single Pier 6 column.

The columns were built in stages. Here iron workers build the second stage of a pier column.

Loading column forms onto a barge. The formwork was stripped after each corresponding stage of construction.

Iron workers at Pier 16 place the steel to tie four rebar columns into one solid column.

These three photos show the sequence of erection for the rebar columns. The rebar column is loaded onto a tip-frame that tilts the column upright so the tower crane can lift it into place. The column is set in place atop the existing column structure—alternating splices tie the two columns together. Iron workers then scale the column, secure it in place, and cut-loose the rigging from the column.

Completed Pier 6 column with
falsework in place to support the
pier-table construction. The rebar
and forms for the stem and
diaphragm walls are in place.

Pier Tables:

The bridge was constructed with
one pier table over land and ten
pier tables over water. The pier
tables connect the tops of the pier
columns to the segmental bridge
deck. All eleven of the pier tables
differ slightly in their design; how-
ever, the same general procedures
were used to construct them. All
together the pier tables contain
2,498 tons of rebar and over 11,000
cubic yards of concrete. Nearly
100,000 man-hours were spent over
a 2 year period constructing the
pier tables.

The pier tables were built on falsework consisting of steel pipes, steel
beams, and work decks surrounding
the columns. Once the falsework

was in place the formwork was constructed, rebar installed, and the concrete poured. The pier table was poured in three different stages. The bottom deck was poured first, then the stems and diaphragms, and finally the top deck.

Looking north from Pier 5 in October of 2005 we see completed segmental portion reaching out towards Pier 6. Notice in the background the blue form traveler already in place at Pier 7.

CHAPTER EIGHT

Form Travelers and Segmental Construction

The most visually arresting aspect of the project was the use of the traveling form system to build the cantilevered road deck between the piers. For over two years commuters on the existing bridge, who otherwise could see very little of the construction, were treated to an engineering *tour de force* as the graceful curve of the road deck began to take shape. The traveling formwork for the segmental portion of the bridge required a complex system of self-advancing steel forms and a top deck support system.

The formwork and the upper works were assembled and fabricated in the Vallejo yard. They were transported to the job via barge and erected onto the pier tables. In total, there were 4 pairs of travelers (1 pair per cantilever). The multiple sets permitted the simultaneous construction of 4 cantilevers.

It took only five days to erect the two 125 ton travelers onto a pier table. The travelers were erected 22 times over the duration of the project. The accompanying photographs illustrate some of the details of this interesting aspect of the project.

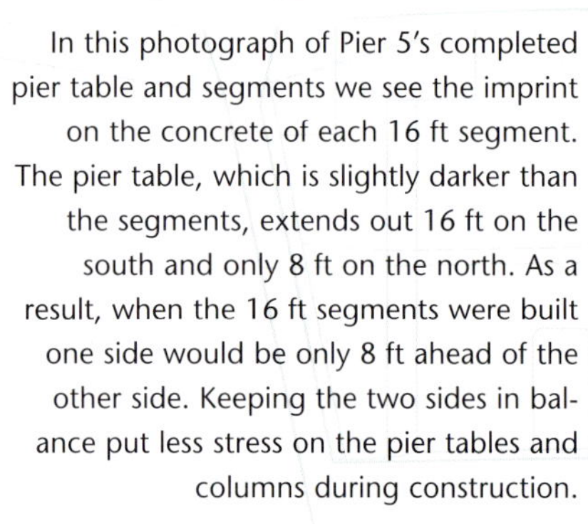

In this photograph of Pier 5's completed pier table and segments we see the imprint on the concrete of each 16 ft segment. The pier table, which is slightly darker than the segments, extends out 16 ft on the south and only 8 ft on the north. As a result, when the 16 ft segments were built one side would be only 8 ft ahead of the other side. Keeping the two sides in balance put less stress on the pier tables and columns during construction.

This photograph of the cantilever segments reaching out from Pier 5 shows the curve as each segment gets shallower. Segments reaching out from Pier 6 will meet these at the midway point between the two piers. The bridge is essentially a hollow, concrete box girder. It is possible to walk from shore to shore inside the bridge deck.

This Photograph from June of 2006 shows the Pier 6 form traveler moving south towards Pier 5. This view shows the adjustable side-panels which allowed the forms to accommodate the changing depth of the segments.

A pump truck on a barge was used to get concrete from water level to the road deck.

Interior panels of Pier 9 looking south towards Pier 8 and the Martinez shore.

Looking north at the traveler closing the gap between Pier 8 and Pier 7.
Below the crew boat *Provider* moves past Pier 8.

Form traveler is free of Pier 7 and ready to move forward to begin the next cycle of construction.

Detail of form traveler at the completion of its work between two piers.

Detail of the face of a segment. The tubes visible in the segment are post-tensioning ducts. As each segment was completed, steel tendons were inserted into the ducts. The ends were anchored and then the tendons were pulled tight with a hydraulic ram. Post tensioning strands were used to tie the segments together, and to the pier tables. The post-tensioning added strength to the entire structure and made the segments a continuous unit. Engineers measured the hydraulic pressure and the tendon elongation to ensure each was stressed to the proper tension.

Iron worker moves the post-tensioning ram into position at the edge of a segment.

Worker on a temporary work platform on Pier 5. The trestle leading from the shore to Pier 6 is visible below.

The bottom deck of a segment is poured. The bottom deck stems, and top deck stems were poured together in one day.

Workers inside traveler at Pier 9.

Removing the side panels of the form traveler at Pier 8. Each time the traveler was removed from a segment the components were shipped to the Vallejo yard for cleaning and were then transported back to the job to be erected at a new pier table.

CHAPTER NINE

The Hinges

The Benicia-Martinez Bridge contains five structural hinges designated A through E. These hinges allow the structure to expand and contract due to temperature changes and move during earthquakes. The hinges separate each structural frame of the bridge. Hinges A, B, and E are concrete box girder structural hinges, with concrete elements and steel sliding bearings. Hinges C and D also included structural steel girders to transfer load across the hinge. All of the hinges are finished off with a rubber and steel expansion joint at the roadway surface.

Hinge E is located between Piers 4 and 5 above Martinez shore. The hinge was formed by a concrete "Upper" and "Lower" hinge seat that has a sliding bearing sandwiched between the two seats.

Casting the lower hinge seat was performed on conventional pipe falsework during the first phase of the project. Work on the upper portion of the hinge had to wait until the cantilever segments were constructed south from Pier 5.

By October of 2005 falsework was in place and workers were ready to construct the upper part of Hinge E that rests on sliding bearings and sit over the lower seat that juts out from left of frame. The Manitowoc 4100 crane in the photo hoisted the 40-ton platform into place.

After the concrete work was complete on the upper portion of the hinge, the falsework stayed in place while post tensioning and dry finish work were completed.

Hinges C and D

Hinges C and D are identical structures, with Hinge C located at Span 12 between Piers 12 and 13 and Hinge D located at Span 8 between Piers 8 and 9.   Each of these hinges is made up of three concrete segments, three structural steel beams, and ten steel sliding bearings. Because these hinges are constructed at the tips of two cantilevers, the sequence of construction was tightly controlled to avoid overloading or damaging the partially finished bridge.

Two views of the hinge gap at span 12. Looking up from the pier table and looking north from the road deck. A temporary steel footbridge was installed to allow workers to safely traverse the gap.

Raising the Hinge C falsework platform into place at span 12. (June 2006)

Four strand jacks were used to lift the platform off the barge and into position at the hinge. The jacks were set in frames on the bridge deck, and the strands were lowered through holes in the deck. The weight lifted was approximately 450 tons. It took about four hours to raise the platform into place. When the time came to remove the platform, the same jacks lowered the platform back onto the barge.

Here we see one of the four strand-jacks used to lift the hinge falsework into place.

The hinge falsework platform was designed to span across the gap at the hinge (initially 60 feet wide) and hang from both cantilevers. The platform was designed with bearings that allowed rotation. One end was also locked to prevent longitudinal movement. This allowed the platform to be locked at one end but not the other, preventing stress buildup in the platform.

Hinge C secured in place. It is adorned with a Kiewit flag and an American flag in the manner of a "topping out"—a time honored tradition in the construction industry.

Looking down from the road deck at the three massive girders that support the hinge joint.

This view from inside the bridge shows the enormous size of the hinge girders.

This aerial view from July 2006 shows work progressing on Hinge C.

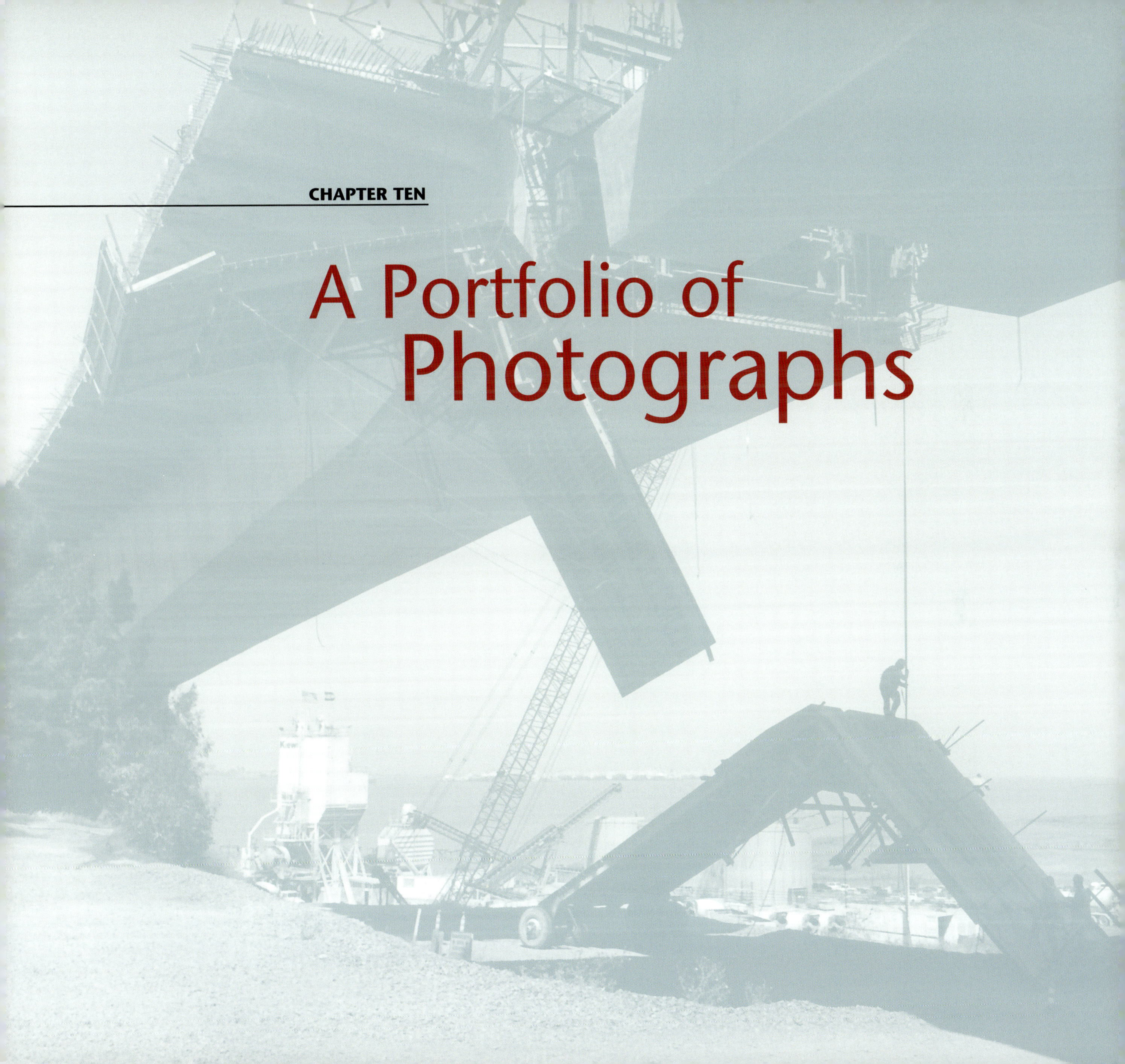

A Portfolio of Photographs

This aerial photo from April 2006 looking from Martinez to the Benicia shore provides a splendid view of the two existing bridges as well as the progress of the new bridge. (Photo by heliphotos@mchsi.com)

The American flag waves proudly at the entrance to the south trestle with Pier 6 in the background.

The *Global Pioneer* moves under the lift span while a tug waits to bring a barge of concrete trucks back to the south trestle.

Night view of Piers 7 through 14
taken from the south trestle.

Crew boat *Provider*
at twilight.

The old and the new. A view of the new bridge from beneath the old train bridge.

Looking north from the foot of Pier 8 we see the completed segments of Pier 9 waiting to meet with Pier 8. Hinge D would close the gap between the two piers.

Crew boat makes its way past Pier 9 on the way to the north trestle.

This view from the top of Pier 8 looks north at the completed segments of pier 9. The various holes are for the post-tensioning strands that tie all the segments together and the cutouts are for steel girders that will make-up hinge D.

Form travelers in place during the early stages of segmental construction of Pier 7.

Crane setting a rebar pile-cage into a pile casing.

Detail of tower crane's hook.

Detail of crawler crane's track on top of Pier 9.

Detail of crane boom and jib on north trestle.

A wide-angle view of the project from the stair tower leading up to top of Pier 5.

Concrete truck at the nitrogen cooling station.

Hydro-lift platforms were used to perform concrete detail and finish work beneath the road-deck.

Iron worker tying rebar.

Some cranes were mounted on barges
so they could easily be moved around
the site as needed.

Buoy and anchor on barge.

Detail of hydraulic
equipment.

Tower crane at Pier 9 is removed and loaded onto a barge.

A view the complicated falsework structure that supported Frame 1 at the Benicia shore where the new bridge ties into the I680/I780 interchange.

Looking down from the road-deck at the removal of steel piles at south trestle.

Crane operator in the cab of Pier 14 tower crane. Iron worker is watching as the jib is rigged for lowering to a barge, January 2007.

Rigging the jib during the demobilization of Pier 14 tower crane, January 2007.

Workers loading rigging onto a crew boat at Pier 14.

Crew boat *Hawk* coming in to the north trestle.

Worker loading grout into a mixing machine.

Catamaran barge picks a 1,800 ton stage 1 footing. The "cat–barge" consisted of two 178 ton trusses that spanned two 180 ft x 54 ft barges. Each barge had 2 two-drum hoisting systems that each held 1,800 ft of inch and three-quarter wire rope that threaded through four 500-ton capacity load blocks."

Tower cranes used a jacking ring around the tower to take out tower sections and lower themselves down. Once the crane jib was below the level of the road deck a conventional barge crane took it the rest of the way down. Also shown in the foreground are Kiewit's edge beam forms staged neatly for use in the next location.

The *Terri L. Brusco,* one of the many tugs that moved barges around the site.

Tug Boat *Amy Elise*, pushing a barge of materials, moves past a crew boat heading towards the south trestle.

Another view from Pier 5 looking north across the bay to Benicia.

A view of the beautiful new toll plaza on the Martinez side.

Bibliography

"Benicia- Martinez Bridge: Partnering helps team overcome unexpected challenges." 2005. *Kieways*. 61.2 (April-June): 15-17.

"Benicia-Martinez Bridge Souvenir Edition." 1962. *Contra Costa Gazette*. September 23.

Crane, Veronica and John V. Robinson. 2007. *Port Costa*. San Francisco: Arcadia Publishing.

Gafni, Matthias. 2006. "Time Capsule buried in Benicia span of bridge." *Vallejo Times Herald*. January 17.

"Giant Span Commissioned at Ceremony." 1930. *Richmond Daily Independent*. November 1.

Harding, C. R. 1930. "Pier Construction for the S.P. Railroad Bridge Across Suisun Bay." *Engineering News Record*. (January 16): 174-180.

Kirkbride, W.H. 1934. "The Martinez-Benicia Bridge." *ASCE Transactions*. 99: 154-89.

"Martinez-Benicia Bridge Opened for Service." 1930. *Southern Pacific Bulletin*. (November-December): 3+.

Rego, Nilda. 1988. "Rail span completed vital link." *Contra Costa Times*. June 12.

Roak, S.A. 1930. "Designer Says 'Bridge is Safe.'" *Contra Costa Gazette*. October 31.

———. 1931. "The Design and Erection of the Martinez- Benicia Bridge Superstructure." *Engineering News Record*. (June 4): 918-921.

Robinson, John V. 2005. *Al Zampa and the Bay Area Bridges*. San Francisco: Arcadia Publishing.

Rear, George W. 1931. "The Martinez-Benicia Bridge Over Suisun Bay." *Proceedings: The Journal of the Pacific Railway Club*. XV. 8 (November): 1-15.

Shattock, Ken. 2001. "The Story of the Suisun Bay Drawbridge." *The Whale's Tale*. (May): 6.

Snyder, John W. 1989. "The Southern Pacific Builds a Bridge...The Saga of the Martinez—Benicia Bridge." *Railroad History Bulletin 161* (Autumn): 67-93.

Vorderbrueggen, Lisa. 2003. "Bridge team: Bubbles protect fish from noise." Contra Costa Times. February 1.

———. 2003. "'Bubble tree' aids backlog at Benicia Bridge." *San Ramon Valley Times*. February 15.

———. 2003. "Pile driving now completed on the new 1.7 mile Benicia span." *Contra Costa Times*. June 22.